I0488976

BEYOND SCIENCE

THE THEORY OF TIME TRAVEL

"I see a pattern, but my imagination cannot picture the maker of that pattern. I see a clock but I cannot envision the clockmaker. The human mind is unable to conceive of the four dimensions, so how can it conceive of a God, before whom a thousand years and a thousand dimensions are as one?"

(Albert Einstein, Princenton University press, 2000 p.208)

Contents

"The true sign of intelligence is not knowledge but imagination."

(Albert Einstein)

Preface

The book is intended, as far as possible, to give an idea of how time travel can work if it is possible and how it may affect us in theory. It is for those readers who from a general-scientific and philosophical point of view are interested in the theory behind time travel and are not familiar with the mathematical apparatus of theoretical physics. Various other theories of existence of multiverse are also explained in non-mathematical terms.

Einstein said "Imagination is more important than knowledge". So in order for better understanding of the following content we need to set our imaginations free. May the book help you to answer few of the questions that you are looking for and at the end of it leaves you with a better opinion about time travel.

Acknowledgments

I decided to try and write a book on the theory of time travel, to explain how it is possible and what are the different important aspects that we need to understand in relation to time travel.

In this book I have tried to break down the theories of many physicists in simple words so as to understand how their research supports time travel.

Albert Einstein's theory of relativity has been mentioned more than a couple of times because his theory was more than just an inspiration for me. Usually people never really think if time travel is actually possible or no and most of the times when they do think, it is when they watch a particular science fiction movie based on time travel and at the end of it they just think or talk about how awesome it would be if it were actually possible, and this book is especially for those who are curios to understand the real concept behind it. Many a times we try to search on the internet that if time travel is possible and if so how, but there are no specific theories based on it and even though, there are already considerable amount of books which explain time travel and space-time and the theory of time dilation, I feel that none of them really concentrate on just time travel, which I am really fascinated about.

At first, I thought it was impossible to explain what I think about time travel and how I understand it but it became easier by talking about it with my friends. Just by talking and explaining I realized that people are interested but they just need the right question and a person with the right answer. So in this book I will try to help you raise the right question and also try to provide you with a right answer.

"Look deep into nature, and then you will understand everything better."

(Albert Einstein)

INTRODUCTION

People believe in what they see, hear or read from which they decide if they should believe in it or not, but the key in this is to plant an idea into a person's mind and to believe or to not believe is up to us and that's how we develop a conspiracy where an idea divides individuals based on what they believe in.

To be honest I don't believe in anything that I see or read, because now with access to internet anyone can learn anything and make things up. I believe in things only if they make sense to me that is, even if a particular story may not have significant evidence to back it up, I would still believe it to be true if it makes sense to me.

The content of this theory which you are about to read may make sense to some people and for some it may not. However, the objective of the following content of this book is to plant an idea that is to question to what we believe in and to always look deep into nature so that we understand everything better.

PART I

The Theory Of Time Travel

Everything is made of atoms and molecules, including humans. An atom when electronically charged performs abnormal activities. It changes its position from one place to another, without actually covering any distance, in simple words creating a tangle in other atom causing teleportation.

The split theory proves that an atom or any electronically charged nucleus particle when emitted or released from one place affects its other end, which is also proven by former physicist Niel Bohr, he clearly explained how input of one particle changes the output of another particle.

Thus, proving Albert Einstein's theory wrong, it can be said that one particle has an effect on another particle, thus one electronically charged particle can be teleported from one place to another due to Quantum leap that is the tangled effect caused by electronically charged atoms.

"God does not play dice with the universe" said the great Albert Einstein, he said that God does not believe in probability, but Einstein believing in God itself is a probability.

Understanding Quantum Mechanics is not about understanding cosmology, it is about understanding how things work or in other words how universe works.

Universe to me is nothing but a giant Math problem; it is all an equation waiting to be solved. Stephen Hawking once said that "time is everything". He stated that our universe has a starting point and so it will likely have an ending point, later he proved his own theory wrong by the theory of continuous of time that is, time never had a starting point it was always present even before the big bang which created our solar system, thus proving time as constant. So one question arises – Do we exist without time? And the answer for that would be NO. I believe that without time there is nothing and time is everything, time is the only aspect that proves our existence. In general relativity, gravity and space-time are exactly the same by definition. It is impossible, within general relativity, to separate gravitational fields from the fundamentals properties of space-time. Thus, if one thing that we know that can travel through time is gravity.

Now what if there is no God or anything called as luck, even though there are proofs of God's existence as per Greek, Indian and many other mythology, the word God is not properly defined.

I believe that these beings who people refer to as God are actually super beings that is, humans with very high intellectual capacity and who have a complete understanding of our nature and our universe and the way it works. So what if we replace the word God with super beings, the conclusion would be that humans are looking out for themselves. Now let's assume that in future time travel is possible and humans have access to unlimited time through which they can have access to travel in any time era of any universe like one giant physical time dimension where gravity can be used to manipulate situations or communicate between multiverse, now as per Einstein's theory reality is irrelevant since time is relative it can be stretched and manipulated which means everything that is currently happening is irrelevant because in the alternate reality our past, present and future all exists at the same time just out of our reach.

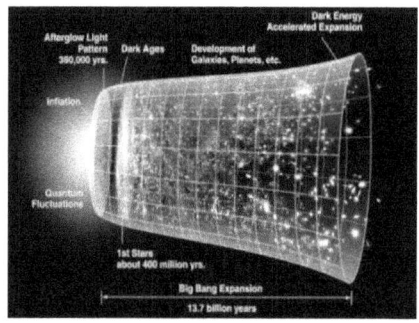

Space-time may stretch out to infinity and if so then everything in our universe is bound to repeat at some point of time and thus creating multiple sets of different universes, where we all may be living alternate reality which is as equal as us living our own reality in our universe.

To prove alternate realities let us take a hypothetical assumption, even though nothing travels as fast as light, it takes approximately 8 minutes and 20 seconds for Sun rays to reach our planet, now consider Sun and Earth as two individual human like bodies, for Sun it has already emitted rays towards Earth but it takes 8 minutes and 20 seconds for the photons emitted from the surface of the sun to travel across the vacuum of space to reach our planet, that means for Sun it has already emitted the photons, but Earth has not yet received therefore we can say that Sun's present is Earth's future. Thus, if we place a giant mirror besides Sun we would be able to see our future 8 minutes and 20 seconds from now.

Now think of it on a large scale where we place a mirror many light years away from Earth, if that were possible we would be able to see the origin or the possible future of human civilization. If we consider all this as a reality we all are dead, alive, not yet born, all at the same time and according to this theory our Earth is itself a set of multiple Earth's with different time lines which all exists at the same time, at the same place and thus making all this, as in our reality irrelevant.

Alternate reality is not just a word in the books now, it's a reality we live with every day and night. It is considered that a person in his life time sleeps for an average approx of 15-18 years. In a day we don't work for 10 consecutive hours or watch television for 10 consecutive hours but we certainly can sleep for 10 consecutive hours. I believe sleeping is not just about getting some rest, it is also a chance for you to live another reality for few hours and our dreams are not just dreams, our subconscious mind tries to tell us something.

So we learnt this above that time is everything and without it we don't exist. So what if that's the key to the next level? What if the only way to go beyond where we are is to try to manipulate time? And the closest we have right now to an alternate reality is our dreams which is something everyone have in common, many people get the same dreams again and again and that could be an indication to something, why would it occur otherwise. It is said that everything that happens, it happens for a reason. So what if our dreams are trying to tell us something and since we cannot manipulate time we can try and manipulate our dreams by controlling our subconscious mind, which may sound impossible but in fact it isn't! If we manage to get a hold on our subconscious mind, we can control our alternate reality and use it to find answers, and go places where our actual conscious mind wouldn't let us go. One of the first philosophers to question the distinction between reality and dreams was Zhuangzi, a Chinese philosopher from the 4th century BC. He phrased the problem as the well-known *"Butterfly Dream,"* which went as follows:

Once Zhuangzi dreamt that he was a butterfly, a butterfly flitting and fluttering around, happy with himself and doing as he pleased. He didn't know he was Zhuangzi. Suddenly he woke up and there he was, solid and unmistakable Zhuangzi. But he didn't know if he was Zhuangzi who had dreamt that he was a butterfly, or a butterfly dreaming he was Zhuangzi. Between Zhuangzi and a butterfly there must be *some* distinction! This is called *the Transformation of Things.*

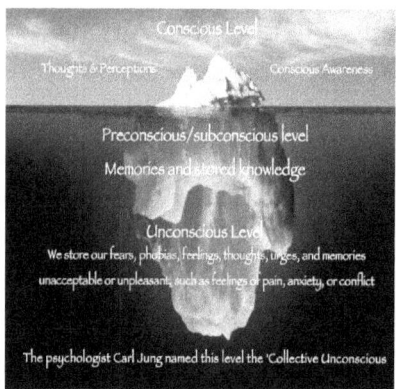

So now we can safely assume that it is possible that alternate realities can exist!

So is inter dimensional travel possible?

I believe that it is not only possible, we might have experienced it ourselves without us actually knowing it and let me explain you how!

In 1957 American physicist Hugh Everett III proposed the many-worlds interpretation of quantum physics, which he termed his "relative state" formulation, which suggested the possibility of all alternate realities existing in our universe which are real within their own worlds. His theory suggests that basically all possible realities in our past or future may or could have happened in another reality. Later His formula was popularized by Bryce Seligman DeWitt in the 1960s and 1970s, there are many theories which can explain travelling in multiverse is possible but his is the only theory which is backed by scientific calculation.

The many worlds interpretation could be one of the many possible ways to resolve the paradoxes that one would expect to arise if time travel actually turns out to be permitted by physics. Entering the past itself would be a quantum event causing breaching, and therefore the timeline accessed by the time traveler simply would be another timeline out of multiple timelines.

(The following diagram explains the same)

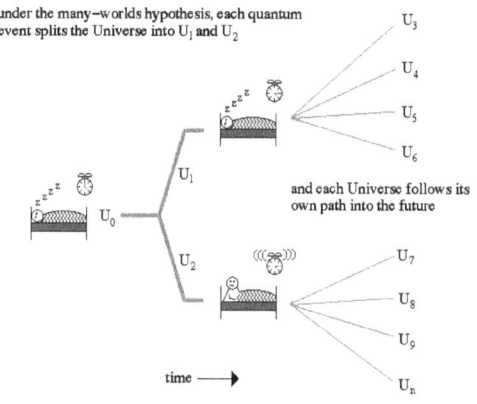

under the many-worlds hypothesis, each quantum event splits the Universe into U_1 and U_2

U_3

U_4

U_5

U_6

and each Universe follows its own path into the future

U_1

U_0

U_2

U_7

U_8

U_9

U_n

time ⟶

all the possible Universe's exist, but none can communicate with another.

WHEN it comes to inter dimensional travel we all can relate to one thing that is déjà vu, which all of us must have experienced at least once in our lives. For those who don't know what déjà vu is, it means a certain situation that we experience, which makes us feel that the specific situation has occurred before and when you think of when it had occurred you forget about it instantly. The concept of "déjà vu" is explained in science as misfiring of the brain and a constant occurrence of this experience makes you feel like you are trapped in a time loop. This is an area which requires a lot more study because no specific reason has proved of why this phenomenon occurs. Then there is, "Jamais vu" which is not as popular as "déjà vu" but in simple terms it means "never seen". It is a feeling where a person recognizes a situation which may have happened but is still a bit unfamiliar with.

As there are no specific explanation of why we experience this, then them being an actual memory from other versions of ourselves which exist in different parallel universes is a possibility, and due to some reason for a second we get an abnormal feeling or an insight of another multiverse where similar situation may have already occurred and so you feel that it happened before but in reality the "other you" had experienced it, but now due to some glitch the mind is inter connected and it is sharing the data and so you feel that you already experienced it. It is a possibility that we could be living in another reality there are people who have had experiences where one day they wake up and everything seems different, not that you are a different person altogether, you are still yourself but things around you have changed, the life you actually lived is now the other yourselves reality and their reality is now your own reality, it is like jumping in some other timeline.

So now, since we have understood the concept of multiverse and alternate realities and how it is possible to actually try to connect to other reality another question arises that can we change our past or our future? According to me NO, you can change the direction to reach the destination but the destination will remain the same. In simple words the outcome will always remain the same, only the situation will differ but that does not mean that our universe is perfect, there is always a glitch in the system, even our world can make a mistake. To explain in simple words let me give you a hypothetical example.

Let's assume that time travel is possible and people travel through time as if it's like any other regular daily activity, now let's assume that you are giving an exam and you don't know what the answer to a certain question is and at that particular moment your future self travels back to the time when you were stuck with that question and your future self gives you the answer to that question and so you solve it and clear the exam and in order to complete the cycle, when you reach at your future selves age you need to travel back in time, when you were stuck with that one question to which you dint know the answer to and give yourself the solution so it completes the circle. So in this entire process who exactly solved the question? NO ONE did, that's right! No one actually solved it, because for the future you the past you solved it and you yourself provide an answer to the past you. This, example explains what a quantum loop is.

Quantum loop is a theory of quantum space-time, according to general relativity, gravity is a manifestation of the geometry of space-time, so if we try to travel in time there is no single objective answer as to how much time has really passed between the departure and return, but there is an objective answer to how much time has been experienced by both the Earth and the traveler, however many in the scientific community believe that backward time travel is highly unlikely, because what if you travel back in time and kill your own grandfather before your father was conceived, then in that universe you wouldn't exist, however this paradox can be avoided due to the

Many worlds theory, which states that irrespective of the outcome, in other multiverse you do exist and what happens in one universe will not have a direct impact on other parallel multiverse.

So travelling back in time may not be possible but theoretically it wouldn't be much of a problem cause even if you die in one universe you do exist in another parallel multiverse.

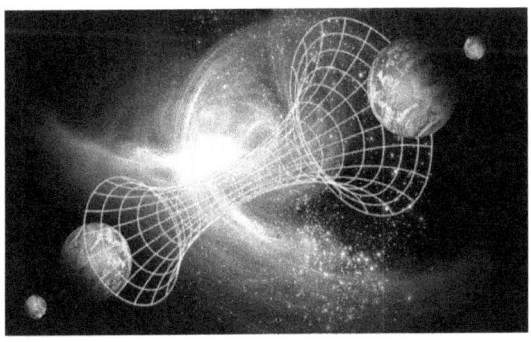

NOW knowing that there are other multiple universes out there, one may think how different would it be from our universe in terms of people living or in terms of technology or atmosphere and climatic condition etc. there are a lot of things that would be different in other universes but if we consider only our parallel universe where multiple Earths exists with different time lines not much would differ. Most of us wanted to do so many things and wanted to achieve so much in life but for various reasons we couldn't, I believe that all these things as in our failed attempts on many things could be an actual reality in some other time line, for instance, since you were a little kid you wanted to become a doctor, but you got less marks and you couldn't get into science and studied commerce instead but in some other reality you could actually be a doctor, in theory it does make sense because for everything that happens, there has to be a reason for it to happen, so what about our desires and hopes which we could not achieve in our universe but still it does not explain why we had that desire or that hope in the first place, the only way it can be reasoned is knowing that in other universe you did accomplish on our dreams and desires. Even though not much would differ we still would be different people living different lives which are real in their own universe as in for us their universe would be different and for them ours would be different so all these universes if exist are real in their own world, like a mirror image of our universe, which are somehow interlinked.

So in simple words we all might be having many different versions of ourselves where we might be living our dreams or we might be having worse life as compared to our real lives back on our original universe.

So now theoretically we don't have to be disappointed when we have a dream or a desire which for some reasons we could not achieve or fulfill in this universe because there is a possibility that we might have in another universe, but this does not mean we should give up on them. By giving up on a dream you will also give up on a possible reality.

Wormholes

A wormhole is a hypothetical tunnel which connects two different points in spacetime.

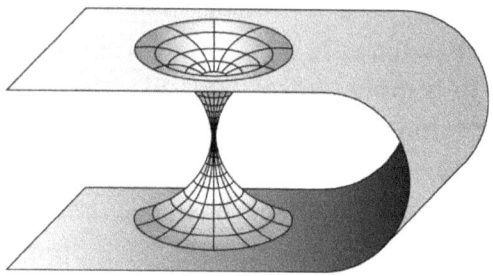

Wormhole may connect a universe with itself which provides a means of interstellar travel and since it connects two different time eras it can be considered as one of the means of time travel. Wormholes may also be connected to infinite series of parallel universes. According to previous studies, the *CASIMIR EFFECT* shows that, quantum field theory allows the energy density in certain regions of space to be negative relative to ordinary vacuum energy and therefore such an effect in space can make it possible to stabilize a traversable wormhole. A traversable wormhole is a type of a wormhole with enough negative energy to stabilize it. The studies have shown that tiny wormholes may appear and disappear spontaneously and with enough negative energy we could stabilize the wormhole theoretically. Thus, if traversable wormhole can be contained it would alter the speed of time and allow us to travel in time. However, while passing through the tunnel of a wormhole the time for the traveler would change drastically when compared to Earthlings.

How gravity plays an important role in the theory of time travel.

In Newton's theory, gravity makes particles leave their straight paths, Newton's law of universal gravitation states that every mass attracts every other mass in the universe and the gravitational force between the two bodies remains proportional to the product of the masses and inversely proportional to the square of the distance between them, what relevant is that in simple words his theory stated that without outside force two stationery objects will never move whereas, In Einstein's theory of general relativity, gravity is a distortion of space-time, which makes time relative and since it has an inverse relation with gravity it is implied that, where the gravitational force is strong, time runs slow and where the gravitational force is weak time runs faster, so the relativity of time depends upon the gravitational force. Any event that occurs in the universe has to involve both space and time and gravity does not just pull space it also pulls time.

Gravity is the manifestation of the geometry of space-time thus, the curvature of space-time not only stretches but also shrinks the distance depending upon the direction and the force of gravity and slows down the flow of time this effect is also called as gravitational time dilation. Thus, the deeper we travel in space against sun the gravitational force for us reduces and time for us slows down as compared to others here on Earth and therefore, astronauts age fractions of fractions of fractions of a second less than earthlings because we age at a different rate depending on the force of gravity.

So is there a way to manipulate gravity here on Earth itself?

As of now no such technology exists which can produce or deflect gravitational force, but if in future if such a device is created, then it would allow physicists to actively study gravity for the first time, and really test out Einstein's general theory of relativity.

The difference between both the theories is that Newtonian gravity separates space and time and can be written as space-time curvature, whereas Einstein's theory of gravity considers space and time as one aspect ie; space-time and in relevance to time travel Einstein's theory is more supportive, because Newton's theory did not explain the origin of gravity and the reason for why Einstein's theory supports time travel is that his theory proposed that gravity is not a force but an effect of space itself, he said that gravity changes the space around it causing space to be curved and thus as we travel away from the gravitational force its pull decreases which leads to decrease in the travelers weight and slows down the time, which is why I explained earlier that the further we go away from the respective mass its gravitational pull decreases. An example for that would be our satellites. The reason it revolves around our planet or the reason why we revolve around the Sun is because our space is curved, because of the weight of the bodies like the weight of sun which results in gravitational pull, due to which we rotate around it and the increase in the weight will decrease the distance between two bodies thus, if sun's weight increases then, our planet and all other bodies surrounding it will get closer to the sun and vice-versa, thus when satellites get closer to Earth's orbit its gravitational pull increases and the satellite falls with increasing velocity and simultaneously its weight increases as it falls from the outer space.

So we can conclude by this that greater the mass, the greater will be the gravitational pull and as we try to go further and further away from that mass, the gravitational pull decreases thus, proving gravity not as a force but just an effect of any mass on space itself.

(The following diagram will help you understand the same)

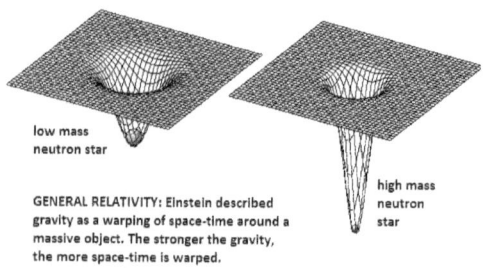

low mass
neutron star

GENERAL RELATIVITY: Einstein described
gravity as a warping of space-time around a
massive object. The stronger the gravity,
the more space-time is warped.

high mass
neutron
star

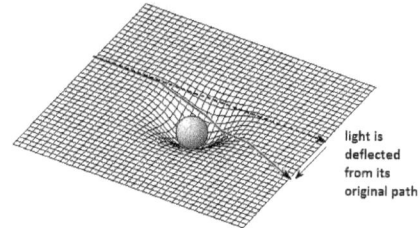

light is
deflected
from its
original path

GENERAL RELATIVITY: Light travels along the curved space taking the
shortest path between two points. Therefore, light is deflected
toward a massive object! The stronger the local gravity is, the
greater the light path is bent.

As we can clearly see in the diagram above that how low/high mass star creates a curve in space and depending upon the mass the curve is bigger or smaller and in the third diagram we can see that because of the curve which the weight of the star creates, the light is deflected from its original straight path.

So it is not just light but time also has an effect from gravity, previously I explained how if we travel against the pull of gravity the time for us slows down, now combining this theory with the theory of curvature of light in the presence of gravity, we can conclude that time also has an effect from gravity, thus at the point of singularity the time we know cease to exist and all multiple realities exists at the same time and time for the space travelers changes drastically than compared to Earthlings.

To conclude this, it would be wise to say that time and light both are affected equally by the effects of gravity. One may think that when at a particular point all realities exist at the same time then what happens to "Now"? And this concept, of what happens to "Now" as in our present time, even troubled Albert Einstein.

He explained that, the experience of the "Now" means something special for man, something essentially different from the past and the future, but this important difference does not and cannot occur within physics so he concluded in his research *NOW: THE PHYSICS OF TIME*, that "there is something essential about the Now, which is just outside the realm of science". Some physicists therefore concluded that the concept of the Now is an illusion. For many reasons it does make sense because having multiple as well as alternate realities at the same time makes the concept of "Now" irrelevant. Even though it does make sense I do believe that they have it all backwards, because if we conclude by stating that the concept of the Now is an illusion then there is no future. To me, the concept of the Now is one reality out of many other possible realities that we live in. The key for this basic understanding is that the objective of studying physics is to understand and explain reality and not deny it, and by stating the concept of Now as an illusion we are not only denying our reality we are also denying a possible future which may affect other alternate realities like ripple effect. For the readers who are not aware of the ripple effect theory it is a situation in which, ripples expand across water when an object is dropped into it, an effect from an initial state can be followed outside gradually. In simple terms a spreading effect or series of consequences caused by a single action or an event. Even a slightest interference can change everything.

There are two ways to understand this, if we travel in time to our past to change an outcome of a certain event, we are not just changing our future we are changing everyone's future. For instance if I travel back in time 10 years ago, when I was in a chess competition and I lost, but now that I know what my opponents move is I can avoid that check mate and now there is a possibility that I might win so by interfering with the past, I not only changed my future but also my opponents future so now there are two realities that exist in the time dimension one where my opponent won the game and the other one, when I won the game, the other reality is created by me so whatever happens in that reality will be only because I decided to change the outcome of that game in chess only by preventing a single check mate and then one thing leading to another the alternate reality created by me will differ from the original reality.

(The following two diagrams will explain the above mentioned illustration)

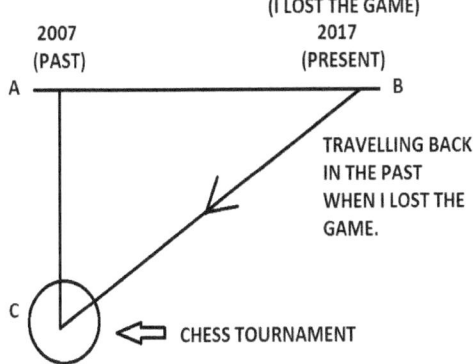

In this diagram I travel back in the past to the time when I lost the game with an objective to change the outcome of the game. As you can see I travel from point B (present) to point C (past). This is the original reality where I lost the game.

As shown in the second diagram, I create an alternate reality from point C to point D by changing the outcome of the game so whatever happened from point A-B will differ from point C-D. So by travelling back in time I not only changed the outcome of the game, I also created a whole new world which is different than our original world, by this we can conclude that every single step that we take or every single decision that we make, will have an effect not just in our reality but also in another reality.

Now a question arises that, if time travel is possible in the near future then how can we know that if our future self travelled back in time and has changed the outcome for our betterment? Well, that's where the concept of déjà-vu and Jamais-vu is interlinked. According to me this misfiring of the brain that occurs is a type of an anomaly due to which we think that a particular event has occurred before, but in reality it is a situation that we foresee from our another reality.

(For the readers who are not aware of what an anomaly is, in simple terms it means something that deviates from what is normal.)

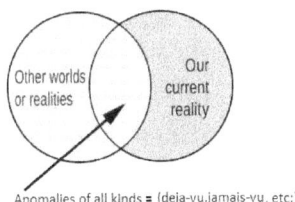

Anomalies of all kinds = (deja-vu,jamais-vu, etc;)

The simple definition of anomaly as stated above means, something that deviates from what is normal that means it can be anything, it can happen to anyone call it a vibe or déjà-vu or a person with psychic abilities and on and on the list continues, our world is not perfect there are loopholes and in the same way anomaly is a kind of a loophole where multiple realities co-exist simultaneously.

This can be done even at the given moment, you can right now create an alternate reality for yourself, for instance your attention is focused on your current reality right now, and your desired situation can be said to exist as some alternate reality outside you're primary focus. In that other reality you already are where you want to be and now in that reality you have different goals and different objectives and so you want something more for yourself so you switch dimensions again and again this continues infinite number of times thus, proving endless number of realities. It may sound odd at first but it is fairly straightforward. There is always a relation between what you vibe or when you have a déjà-vu in a certain situation with the current reality, we always say that everything happens for a reason then why we fail to establish a good reason behind what we dream or our relation with what we vibe or when we have an intuition or when we simply just get lucky in a certain situation.

In a world of science where we can give a scientific reason for anything that exists we still think of a person being lucky or not lucky in a certain situation. There may or may not be any relation but the fact that we can question it, is more than enough reason to believe that a relation between them is possible!

Stephen Hawking in his book *"A BRIEF HISTORY OF TIME", chapter 8 - The origin and fate of the universe*, talks about Albert Einstein's theory of general relativity which predicted that space-time began at the big bang singularity and would come to an end at the big crunch singularity (if the whole universe recollapsed), or at a singularity inside a black hole (if a local region, such a star, were to collapse) and when quantum mechanics is added to this theory the whole equation changes.

I previously explained how time is relative and it has no beginning or an end, and how it is the only constant we know that did exist before us and will continue existing long after we are gone. One can argue by stating that at the point of singularity when our world was created everything had to be very precise in the making, that is the right atmosphere, the right temperature, the right gravitational pull and so on and this is why I mentioned super beings, you can call it a force but clearly it is not god.

Our future selves are looking after us and not god because if you cannot believe in four dimensional world how can you believe in one super natural entity who has the power to control everything we can possibly imagine.

THE CONCEPT OF FACT AND ILLUSION

Is time an illusion? We know that time is the only constant aspect that proves our existence but that is based on a fact that time can be measured. Without measurement we cannot prove the existence of time. In 1978,John Wheeler proposed Delayed Choice Theory, to prove that the result can only be measured at the end of the objects journey. At that time, the technology was not available to conduct a strong experiment to back up the theory, but scientists now have been able to successfully conduct the experiment and proved that the physicist was right.

Physicist Andrew Truscott from the Australian National University (ANU), said: "it proves that measurement is everything. At the quantum level, reality does not exist if you are not looking at it."

We always consider time to always run in forward direction without questioning why?

The laws of physics are symmetric which means that time could have easily moved in a backward direction as it does forward. Einstein's theory of relativity states a four dimensional structure where everything that has happened or is happening or everything that will happen has its own co-ordinates in space-time. This means that our past and future are equally important as present, but this still does not make our present an illusion it is just an idea that since everything exists and has always been existed and will forever exists in space-time the present has lost its credibility.

When we talk about a beginning and an ending point, the idea of our universe to be ever expanding doesn't add up. Yes it is increasing and may keep on increasing forever but at big crunch all that we know will be destroyed creating new universes, new planets, etc. so it still may expand but we might not experience it and since we can't then theoretically for us the universe is finite but technically and logically this process is never ending. The big crunch will result into the big bang which created our universe and then will create another universe and thus this will go on and on until infinity.

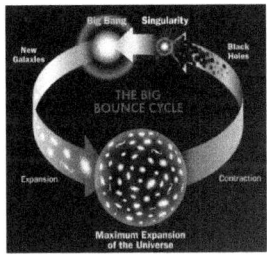

Thus, to think of time as an illusion may not be wrong but it may shatter the fundamentals of physics on which all our theories are based.

Earlier many theoretical physicists believed that time is nothing but a product of our consciousness and that our mind records what we have seen in the past and can also retrieve it and from our daily experiences, we cannot construct what is going to happen in the future but in the future we can retrieve that moment and thus this theory defined our past, present and future.

According to me time is not an illusion, even though it is logically finite for us but time as a whole is for infinity, irrespective to our existence in perspective and time will go on even without us and so it cannot be considered as an illusion.

The importance of relativity is that there is no absolute time and no absolute space and everything is relative and in order for better understanding of time we need to separate it from the evolution of time.

SIMULATED REALITY

Simulated reality is a theory which states that our reality could be simulated. In other words it means that we may or may not be living inside a simulation. Now given the current state of technology nothing close to this is achievable. However, as discussed earlier I stated that its not God and super humans that is our future selves are looking out for us and in this case they are controlling our conscious mind.

The problem with simulation is that, it is indistinguishable from the true reality, that means unless we can prove the existence of simulated reality, we can never be sure if we are living in one right now or no.

Even though we haven't reached a level of technological maturity where we can achieve this, but in the future due to technological advancement we surely can achieve it. So we cannot completely rule out this hypothesis.

One can argue by stating that simulated objects cannot satisfy physical bodies and that our brain is not a machine, but the complete idea of simulation is based on controlling the conscious mind and make us believe that what we see and what we feel is real in other words the hypothesis involves real objects, real people with real conscious but our conscious mind is itself not in our control, so we can't tell the difference between what we think is real and what actually is real.

Many may get confused between virtual reality and simulated reality. However, both are very different from each other. Virtual reality is a system that enables a person to react and move according to certain simulated conditions whereas simulation is a program which models a real life situation by putting values into a model to see how it behaves or reacts given a particular situation, which in our case as per the hypothesis that model is us.

Virtual reality games are gaining more and more popularity and simple games are becoming more and more complex and indistinguishable from reality. What if in future we add more than just colors on a screen, what if we add smell and taste and the ability to feel whatever we see on the screen, with the advancement in technology the possibilities are endless. There are many reasons to believe that our universe might be a simulation and one of them is that in order for anything to qualify as a simulation it has to be finite and everything around us is finite, even though time itself is for infinity for us human beings, it is finite. This property makes our universe both computable and limited and thus, it qualifies as a simulation.

So in conclusion to simulated reality, the fact that we cannot prove it to be wrong and we can question its existence is more than enough a reason to believe that it could be possible that in post human stage, civilization with advanced computing power might decide to run a simulation of their ancestors in the universe.

CONCLUSION

Most of the scientists in the world are too occupied in developing new technology and new theories that explains the "WHAT" part of the universe instead of researching on the "WHY" part. And for the people who are more fascinated about the "WHY" part, the philosophers are not able to keep up with the present advance theoretical physics, and so in time, the number of people who think about the "WHY" part are decreasing and what important is that at the end of it, we all want to make sense of what we see around us and to think about what our place is in the universe and why is it the way everything is? And to understand all this we all need to develop a basic understanding on life.

We all have something to learn, we all need to perform our duties first as humans. According to me, we humans have created an illusion for ourselves which we refer as normal.

I believe that, we were not meant to do things that we are currently doing, human beings are capable of so much more that it makes me believe that what we do has no real impact on our nature. The lives that we live, the jobs that we do and many other things are just an illusion where as we have yet to discover reality, if we live our own lives and not think anything beyond that, we will never actually evolve into greater beings. By just doing what we do, we are limiting our own potential.

How is this even possible that out of billions of planets in our ever expanding space only one planet has life, with at least 200 billion galaxies out there (and possibly more), we're very likely talking about a Universe filled with septillion number of planets and that to accounted only in our observable universe, so even if there is a slightest possibility that we are not alone is more than enough a reason for us to evolve and do things that actually matter and have an impact in real life, Just because we cannot see anything doesn't mean things are not changing, in reality everything is constantly changing we only need to broaden our minds and think beyond our desires and only then we could begin being an evolved being.

It seems like In time humans have stopped trying to evolve, it seems that we are only trying to get comfortable with the situations that are thrown at us and live with the same consequences and we let our jobs define us.

For instance everyone introduce themselves as doctors, lawyers or an accountant etc. but why do we stop there why can't we be more than just our profession or a job that defines us. If we try to get out of this illusion, we will be able to see that real life is way more beautiful than our normal lives.

We all need to start doing things that have an impact on our real life or on our nature and not just on our profession or on any other material thing.

Our universe is so amazing that no matter how much you learn about it, there will always be something that you didn't know.

"Learn from yesterday, live for today, hope for tomorrow. The important thing is not to stop questioning."

(Albert Einstein)

DEFINITIONS

- <u>ATOM</u>- The basic unit of ordinary matter, made up of a tiny nucleus (consisting of protons and neutrons) surrounded by orbiting electrons.

- <u>SINGULARITY</u>- A point in space-time at which the space-time curvature becomes infinite.

- <u>SINGULARITY THEOREM</u>- A theorem that shows that a

singularity must exist under certain circumstances – in particular, that the universe must have started with a singularity.

- <u>SPLIT THEORY</u>- Photons or particles of matter (like an electron) produce a wave pattern when two slits are used.

- <u>QUANTUM</u>- The indivisible unit in which waves may be emitted or absorbed.

- <u>QUANTUM MECHANICS</u>- The branch of mechanics that deals with the mathematical description of the motion and interaction of subatomic particles, incorporating the concepts of quantization of energy, wave particle duality, the

uncertainty principle, and the
correspondence principle.

- <u>QUANTUM LEAP</u>- A sudden large
  increase or advance.

- <u>PHOTON</u>- A quantum of light.

- <u>WEIGHT</u>- The force exerted on a
  body by a gravitational field. It is
  proportional to, but not the same
  as, its mass.

- <u>BIG BANG</u>- The singularity at the
  beginning of the universe.

- <u>COSMOLOGY</u>- The science of the
  origin and development of the
  universe. Modern cosmology is
  dominated by the Big Bang theory,
  which brings together

observational astronomy and particle physics.

- **GENERAL RELATIVITY**- General Relativity is the geometric theory of gravitation published by Albert Einstein in 1915 and the current description of gravitation in modern physics. (Difference between general and special relativity) – Special relativity doesn't include gravity, whereas general relativity does. In special relativity the laws of physics are the same for all internal coordinate systems; that is those in which Newton's first law of motion is true.

- **GALAXY**- AN enormous cluster of stars, planets, gas and dust.

Galaxies may contain billions of stars.

- <u>SPACE-TIME</u>- The concept of time and three-dimensional space regarded as fused in a four-dimensional continuum.

- <u>SPACETIME CURVATURE</u>- In general relativity it is assumed that space-time is curved by the presence of matter (energy/mass).

- <u>MULTIVERSE</u>- A hypothetical space or realm consisting of a number of universes, of which our own universe is only one.

- <u>ANOMALY</u>- Something that deviates from what is standard, normal, or expected.

- <u>EVENT</u>- A point in space-time, specified by its time and place.

- <u>RIPPLE EFFECT</u>- The continuing and spreading results of an event or action.

- <u>TIME DILATION</u>- Time dilation is a difference of elapsed time between two events as measured by observers either moving relative to each other or differently situated from a gravitational mass or masses.

- <u>BIG CRUNCH</u>- A contraction of the universe to a state of extremely high density and temperature. In simple words (a hypothetical opposite of the big bang).

- **CONSERVATION OF ENERGY- The law of science that states that energy or (its equivalent in mass) can be neither created nor destroyed.**

- **LIGHT SECOND (LIGHT YEAR)- The distance travelled by light in one second (year). One light year is equal to 9.5 trillion kilometers or 5.9 trillion miles.**

- **CASIMIR EFFECT- The Casimir effect is a small attractive force that acts between two close parallel uncharged conducting plates. It is due to quantum vacuum fluctuations of the electromagnetic field. The effect was predicted by the**

Dutch physicist Hendrick Casimir in 1948.

* **INTERSTELLAR TRAVEL-** Interstellar travel is the term used for hypothetical piloted or unpiloted travel between stars or planetary systems. The speed required for interstellar travel in a human lifetime far exceed what current methods of spacecraft propulsion can provide

The past that we lived, the present that we live and the future that we may experience is all forged. It is an edited version of a raw video clip, silently decorated with better words and many undisclosed corners of a dark room.

But why do we care?

After all we became a part of the same system. A system of organized chaos, where people care only about the present, which will soon be a forged past and later, a forgotten History.

-Vaibhav Kamani.

THE END

ISBN-9781520809731